Intermediate 2 | Units 1, 2 & 3

Mathematics

Leckie×Leckie

First exam published in 2002.
Published by Leckie & Leckie, 8 Whitehill Terrace, St. Andrews, Scotland KY16 8RN tel: 01334 475656 fax: 01334 477392
enquiries@leckieandleckie.co.uk www.leckieandleckie.co.uk

ISBN 1-84372-330-1

A CIP Catalogue record for this book is available from the British Library.

Printed in Scotland by Scotprint.

Leckie & Leckie is a division of Granada Learning Limited, part of ITV plc.

Acknowledgements

Leckie & Leckie is grateful to the copyright holders, as credited at the back of the book, for permission to use their material.
Every effort has been made to trace the copyright holders and to obtain their permission for the use of copyright material.
Leckie & Leckie will gladly receive information enabling them to rectify any error or omission in subsequent editions.

[BLANK PAGE]

X100/201

NATIONAL
QUALIFICATIONS
2002

MONDAY, 27 MAY
1.00 PM – 1.45 PM

**MATHEMATICS
INTERMEDIATE 2
Units 1, 2 and 3
Paper 1
(Non-calculator)**

Read carefully

1 **You may <u>NOT</u> use a calculator.**

2 Full credit will be given only where the solution contains appropriate working.

3 Square-ruled paper is provided.

LIB X100/201 6/21570

SCOTTISH
QUALIFICATIONS
AUTHORITY

©

FORMULAE LIST

The roots of $ax^2 + bx + c = 0$ are $x = \dfrac{-b \pm \sqrt{(b^2 - 4ac)}}{2a}$

Sine rule: $\dfrac{a}{\sin A} = \dfrac{b}{\sin B} = \dfrac{c}{\sin C}$

Cosine rule: $a^2 = b^2 + c^2 - 2bc \cos A$ or $\cos A = \dfrac{b^2 + c^2 - a^2}{2bc}$

Area of a triangle: $\text{Area} = \frac{1}{2} ab \sin C$

Volume of a sphere: $\text{Volume} = \frac{4}{3} \pi r^3$

Volume of a cone: $\text{Volume} = \frac{1}{3} \pi r^2 h$

Volume of a cylinder: $\text{Volume} = \pi r^2 h$

Standard deviation: $s = \sqrt{\dfrac{\sum(x - \bar{x})^2}{n-1}} = \sqrt{\dfrac{\sum x^2 - (\sum x)^2 / n}{n-1}}$, where n is the sample size.

ALL questions should be attempted.

Marks

1. In a tournament a group of golfers recorded the following scores.

    74   70   71   73   75   71   73   72

    72   75   71   76   74   72   70   73

    (a) Construct a frequency table from the above data and add a cumulative frequency column.

    2

    (b) What is the probability that a golfer chosen at random from this group recorded a score of less than 72?

    1

2.

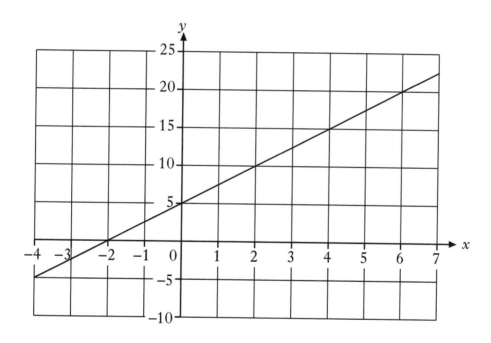

Find the equation of the straight line shown in the diagram.

3

[Turn over

Marks

3.

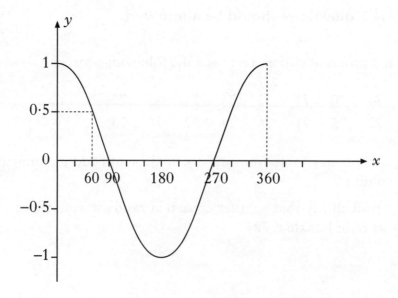

Part of the graph of $y = \cos x°$ is shown above.

If $\cos 60° = 0.5$, state two values for x for which $\cos x° = -0.5$, $0 \le x \le 360$.

2

4. Multiply out the brackets and collect like terms.

$$(x - 3)(x^2 + 4x - 1)$$

3

5. A sample of students was asked how many times each had visited the cinema in the last three months.

The results are shown below.

4	5	4	1	4	3	2	2	4	6	2
3	4	4	1	3	1	2	3	1	1	

(a) From the above data, find the median, the lower quartile and the upper quartile.

3

(b) Construct a boxplot for the data.

2

(c) The same sample of students was asked how many times each had attended a football match in the same three months.

The boxplot below was drawn for this data.

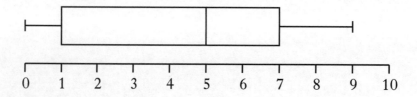

Compare the two boxplots and comment.

1

Marks

6.

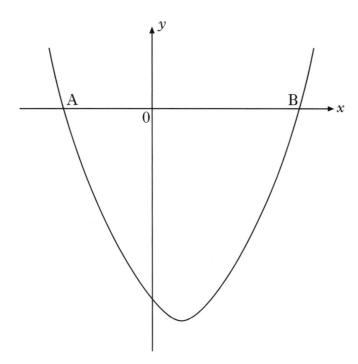

The equation of the parabola in the above diagram is

$$y = (x - 1)^2 - 16.$$

(a) State the coordinates of the minimum turning point of the parabola.   **2**

(b) State the equation of the axis of symmetry of the parabola.   **1**

(c) The parabola cuts the x-axis at A and B. Find the length of AB.   **3**

7. (a) Express $\sqrt{45} - 2\sqrt{5}$ as a surd in its simplest form.   **2**

(b) Express as a fraction in its simplest form

$$\frac{1}{x^2} + \frac{1}{x}, \qquad x \neq 0.$$ **2**

[END OF QUESTION PAPER]

X100/203

NATIONAL
QUALIFICATIONS
2002

MONDAY, 27 MAY
2.05 PM – 3.35 PM

MATHEMATICS
INTERMEDIATE 2
Units 1, 2 and 3
Paper 2

Read carefully

1 **Calculators may be used in this paper.**

2 Full credit will be given only where the solution contains appropriate working.

3 Square-ruled paper is provided.

FORMULAE LIST

The roots of $ax^2 + bx + c = 0$ are $x = \dfrac{-b \pm \sqrt{\left(b^2 - 4ac\right)}}{2a}$

Sine rule: $\dfrac{a}{\sin A} = \dfrac{b}{\sin B} = \dfrac{c}{\sin C}$

Cosine rule: $a^2 = b^2 + c^2 - 2bc\cos A$ or $\cos A = \dfrac{b^2 + c^2 - a^2}{2bc}$

Area of a triangle: $\text{Area} = \dfrac{1}{2}ab\sin C$

Volume of a sphere: $\text{Volume} = \dfrac{4}{3}\pi r^3$

Volume of a cone: $\text{Volume} = \dfrac{1}{3}\pi r^2 h$

Volume of a cylinder: $\text{Volume} = \pi r^2 h$

Standard deviation: $s = \sqrt{\dfrac{\sum(x - \bar{x})^2}{n - 1}} = \sqrt{\dfrac{\sum x^2 - (\sum x)^2 / n}{n - 1}}$, where n is the sample size.

ALL questions should be attempted.

Marks

1. The sketch shows a triangle, ABC.

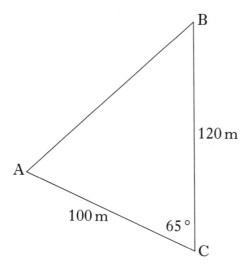

Calculate the area of the triangle. 2

2. Solve **algebraically** the system of equations

$$3x - 2y = 11$$
$$2x + 5y = 1.$$

3

3. (a) The price, in pence, of a carton of milk in six different supermarkets is shown below.

    66   70   89   75   79   59

    Use an appropriate formula to calculate the mean and standard deviation of these prices.
    Show clearly all your working. 4

   (b) In six local shops, the mean price of a carton of milk is 73 pence with a standard deviation of 17·7.

    Compare the supermarket prices with those of the local shops. 2

[Turn over

Marks

4. A pendulum travels along an arc of a circle, centre C.

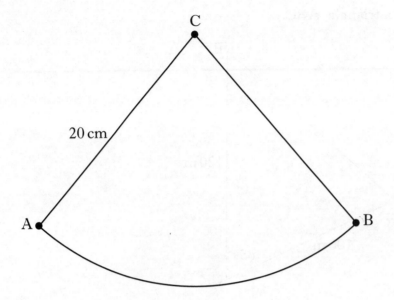

The length of the pendulum is 20 centimetres.

The pendulum swings from A to B.

The length of the arc AB is 28·6 centimetres.

Find the angle through which the pendulum swings from A to B.

4

5. (a) (i) Factorise completely

$$3y^2 - 6y.$$

1

(ii) Factorise

$$y^2 + y - 6.$$

2

(b) Hence express $\dfrac{3y^2 - 6y}{y^2 + y - 6}$ in its simplest form.

2

Marks

6. A container to hold chocolates is in the shape of part of a cone with dimensions as shown below.

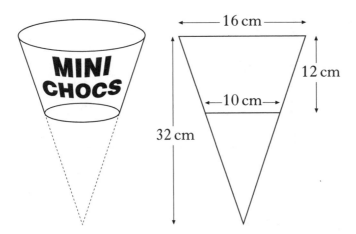

Calculate the volume of the container.

Give your answer correct to one significant figure.   **5**

7. Solve the equation

$$2x^2 + 3x - 1 = 0,$$

giving your answers correct to one decimal place.   **4**

[Turn over

Marks

8. The diagram shows two positions of a surveyor as he views the top of a flagpole.

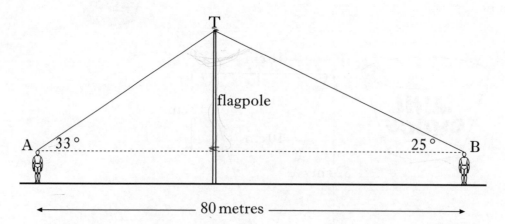

From position A, the angle of elevation to T at the top of the flagpole is 33°.

From position B, the angle of elevation to T at the top of the flagpole is 25°.

The distance AB is 80 metres and the height of the surveyor to eye level is 1·6 metres.

Find the height of the flagpole.

6

Marks

9. The diagram below shows a circular cross-section of a cylindrical oil tank.

In the figure below,

- O represents the centre of the circle
- PQ represents the surface of the oil in the tank
- PQ is 3 metres
- the radius OP is 2·5 metres.

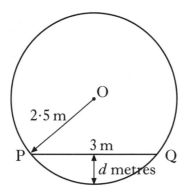

Find the depth, *d* metres, of oil in the tank. **4**

10. The population of Newtown is 50 000.

The population of Newtown is **increasing** at a steady rate of 5% per annum.

The population of Coaltown is 108 000.

The population of Coaltown is **decreasing** at a steady rate of 20% per annum.

How many years will it take until the population of Newtown is greater than the population of Coaltown? **5**

[Turn over for Questions 11 and 12 on *Page eight*

Marks

11. (a) Simplify

$$6x^{\frac{3}{2}} \div 2x^{\frac{1}{2}}.$$

2

(b) Change the subject of the formula

$$r = 3p + 2t$$

to p.

2

12. At the carnival, the height, H metres, of a carriage on the big wheel above the ground is given by the formula

$$H = 10 + 5 \sin t°,$$

t seconds after starting to turn.

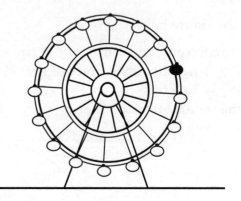

(a) Find the height of the carriage above the ground after 10 seconds.

2

(b) Find the **two** times during the first turn of the wheel when the carriage is 12·5 metres above the ground.

4

[END OF QUESTION PAPER]

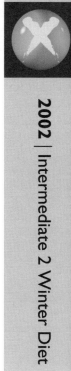

[BLANK PAGE]

W100/201

NATIONAL
QUALIFICATIONS
2002

FRIDAY, 18 JANUARY
9.00 AM – 9.45 AM

MATHEMATICS
INTERMEDIATE 2
Units 1, 2 and 3
Paper 1
(Non-calculator)

Read carefully

1 **You may NOT use a calculator.**

2 Full credit will be given only where the solution contains appropriate working.

3 Square-ruled paper is provided.

SCOTTISH
QUALIFICATIONS
AUTHORITY

FORMULAE LIST

The roots of $ax^2 + bx + c = 0$ are $x = \dfrac{-b \pm \sqrt{\left(b^2 - 4ac\right)}}{2a}$

Sine rule: $\dfrac{a}{\sin A} = \dfrac{b}{\sin B} = \dfrac{c}{\sin C}$

Cosine rule: $a^2 = b^2 + c^2 - 2bc \cos A$ or $\cos A = \dfrac{b^2 + c^2 - a^2}{2bc}$

Area of a triangle: $\text{Area} = \dfrac{1}{2} ab \sin C$

Volume of a sphere: $\text{Volume} = \dfrac{4}{3} \pi r^3$

Volume of a cone: $\text{Volume} = \dfrac{1}{3} \pi r^2 h$

Volume of a cylinder: $\text{Volume} = \pi r^2 h$

Standard deviation: $s = \sqrt{\dfrac{\sum (x - \bar{x})^2}{n - 1}} = \sqrt{\dfrac{\sum x^2 - (\sum x)^2 / n}{n - 1}}$, where n is the sample size.

ALL questions should be attempted.

Marks

1. The marks of a group of students in a class test and in the final exam are shown in the scattergraph below.

    A line of best fit has been drawn.

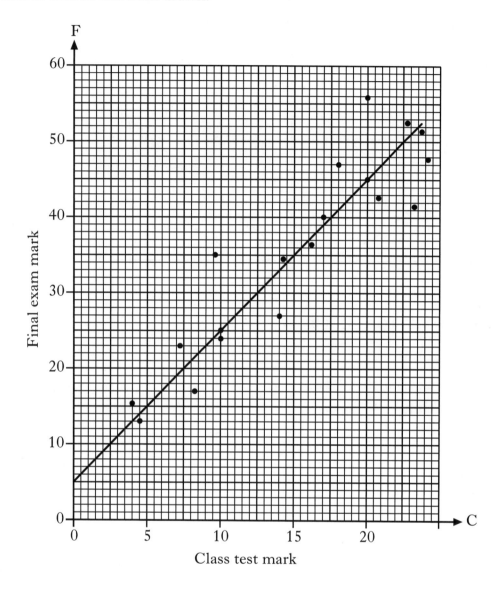

Class test mark

(a) Find the equation of the line of best fit. **3**

(b) **Use your answer to part (a)** to predict the final exam mark for a student who achieved a mark of 12 in the class test. **1**

[Turn over

2.

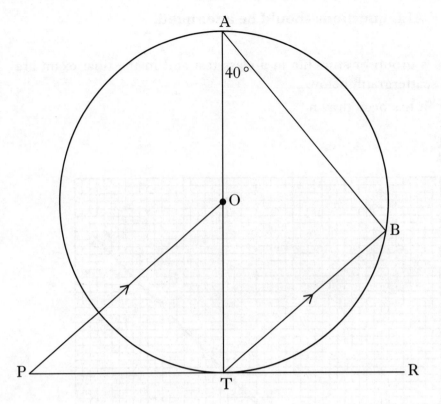

PTR is a tangent to a circle, centre O.

Angle BAT = 40°.

PO is parallel to TB.

Calculate the size of angle OPT.

Show all working.

3

Marks

3. The stem and leaf diagram shows the heights, to the nearest centimetre, of a group of female students.

$$
\begin{array}{r|llllll}
14 & 8 \\
15 & 6 \\
16 & 0 & 4 & 8 & 9 \\
17 & 1 & 2 & 4 & 4 & 5 & 8 \\
18 & 8 \\
\end{array}
$$

$n = 13$ 14|8 represents 148 cm

(a) Using the above information, find

  (i) the median   **1**

  (ii) the lower quartile and the upper quartile.   **2**

(b) Draw a boxplot to illustrate this data.   **2**

(c) A sample of male students from the same course was taken. The heights, to the nearest centimetre, of these students were recorded. The boxplot, shown below, illustrates this new data.

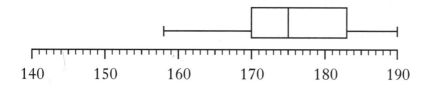

By comparing the boxplots, make **two** appropriate comments about the heights of the female and the male students.   **2**

[Turn over

Marks

4. The diagram below shows the graph of $y = ax^2$.

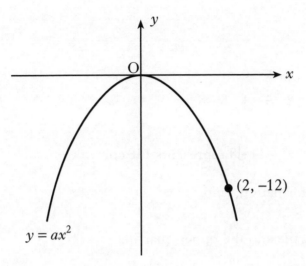

Find the value of a. **2**

5. (a) Express $\dfrac{10}{\sqrt{5}}$ with a rational denominator.

Give your answer in its simplest form. **2**

(b) Evaluate $4a^{\frac{2}{3}}$ when $a = 27$. **2**

6. The graph shown below has an equation of the form $y = \cos(x - b)^\circ$.

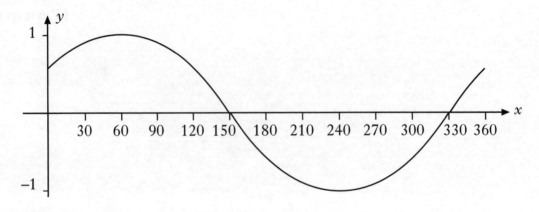

Write down the value of b. **1**

Marks

7.

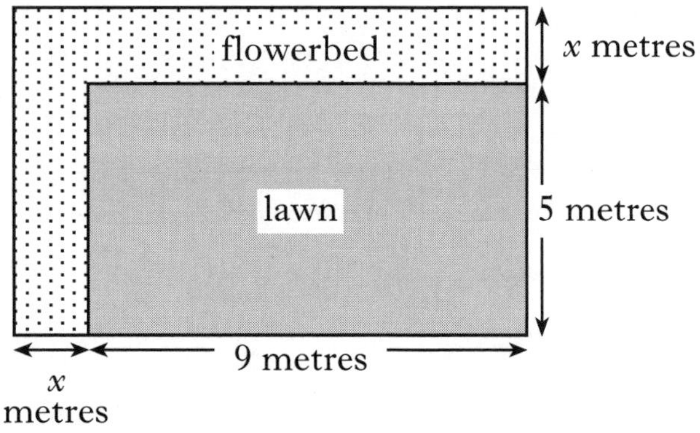

The diagram shows a rectangular garden which consists of a rectangular lawn and a flowerbed along two sides of the lawn

- the lawn measures 9 metres by 5 metres
- the width of the flowerbed is x metres.

(a) State the length and breadth of the garden. **1**

(b) Show that the area, A square metres, of the garden is given by

$$A = x^2 + 14x + 45.$$ **2**

(c) The area of the garden is 77 square metres. Find the width of the flowerbed.

Show clearly all your working. **3**

[END OF QUESTION PAPER]

[BLANK PAGE]

W100/203

NATIONAL
QUALIFICATIONS
2002

FRIDAY, 18 JANUARY
10.05 AM – 11.35 AM

MATHEMATICS
INTERMEDIATE 2
Units 1, 2 and 3
Paper 2

Read carefully

1 **Calculators may be used in this paper.**

2 Full credit will be given only where the solution contains appropriate working.

3 Square-ruled paper is provided.

SCOTTISH
QUALIFICATIONS
AUTHORITY

©

FORMULAE LIST

The roots of $ax^2 + bx + c = 0$ are $x = \dfrac{-b \pm \sqrt{(b^2 - 4ac)}}{2a}$

Sine rule: $\dfrac{a}{\sin A} = \dfrac{b}{\sin B} = \dfrac{c}{\sin C}$

Cosine rule: $a^2 = b^2 + c^2 - 2bc \cos A$ or $\cos A = \dfrac{b^2 + c^2 - a^2}{2bc}$

Area of a triangle: $\text{Area} = \dfrac{1}{2} ab \sin C$

Volume of a sphere: $\text{Volume} = \dfrac{4}{3} \pi r^3$

Volume of a cone: $\text{Volume} = \dfrac{1}{3} \pi r^2 h$

Volume of a cylinder: $\text{Volume} = \pi r^2 h$

Standard deviation: $s = \sqrt{\dfrac{\sum(x - \bar{x})^2}{n-1}} = \sqrt{\dfrac{\sum x^2 - (\sum x)^2 / n}{n-1}}$, where n is the sample size.

ALL questions should be attempted.

Marks

1. Change the subject of the formula

   $$x = y^2 - m$$

   to y.

   2

2. In the diagram opposite AC and BD are arcs of circles with centres at O.

   The radius, OA, is 8 metres and the radius, OB, is 10 metres.

   Angle AOC = 72°.

   Find the shaded area.

   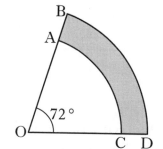

   4

3. The value of a house increased from £85 000 to £86 700 in one year.

   (a) What was the percentage increase?

   1

   (b) If the value of the house continued to rise at this rate, what would its value be after a **further** 3 years?

   Give your answer to the nearest thousand pounds.

   3

   [Turn over

4. A grain store is in the shape of a cylinder with a hemisphere on top as shown in the diagram.

   The cylinder has radius 2·4 metres and height 9·5 metres.

   Find the volume of the grain store.

   Give your answer in cubic metres, correct to 1 significant figure.

   4

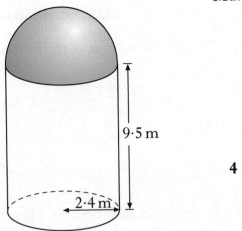

9·5 m

2·4 m

5. At an amusement park, the Green family buy 3 tickets for the ghost train and 2 tickets for the sky ride. The total cost is £8·60.

   (a) Let x pounds be the cost of a ticket for the ghost train and y pounds be the cost of a ticket for the sky ride.

   Write down an equation in x and y which satisfies the above condition.

   1

   (b) The Black family bought 5 tickets for the ghost train and 3 tickets for the sky ride at the same amusement park. The total cost was £13·60.

   Write down a second equation in x and y which satisfies this condition.

   1

   (c) Find the cost of a ticket for the ghost train and the cost of a ticket for the sky ride.

   4

Marks

6. Harry records the amount, in pounds, he earned from his part-time job each week for ten weeks.

14   18   19   20   17   19   18   20   15   22

He calculates that

$$\sum x = 182 \qquad \text{and} \qquad \sum x^2 = 3364$$

where x is the amount in pounds he earned each week.

  (*a*) Calculate the mean amount he earned per week.   **1**

  (*b*) Using an appropriate formula, calculate the standard deviation.   **2**

  (*c*) Irene and Harry compare their earnings over the ten week period. For each of the ten weeks, Irene earns exactly £5 more than Harry.

    State:

    (i) the mean amount Irene earned per week;   **1**

    (ii) the standard deviation of Irene's earnings.   **1**

7. A field with sides measuring 12·5 metres, 13·2 metres and 10·7 metres is represented by the triangle PQR shown below.

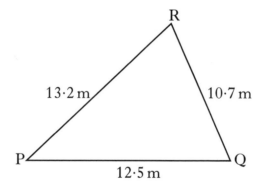

  (*a*) Calculate the size of angle PQR.

    Do not use a scale drawing.   **3**

  (*b*) Calculate the area of the field.   **2**

[Turn over

Marks

8. Solve the equation

$$2p^2 - 3p - 1 = 0,$$

giving the roots correct to 1 decimal place.

4

9. To calculate the height of a cliff, a surveyor measures the angle of elevation at two positions A and B as shown in the diagram below.

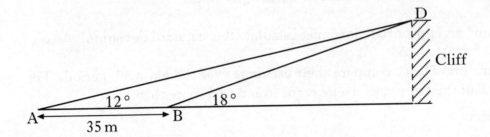

At A, the angle of elevation to D, the top of the cliff, is $12°$.

At B, the angle of elevation to D is $18°$.

AB is 35 metres.

Calculate the height of the cliff.

5

10. (*a*) Multiply out the brackets and collect like terms.

$$(2x + 3)(x^2 - 5x + 2)$$

3

(*b*) Factorise

$$2x^2 - 7x - 9.$$

2

11. Solve the equation

$$2 \tan x° + 4 = 5, \quad 0 \le x < 360.$$

3

Marks

12. (*a*) Simplify

$$\frac{b^{\frac{5}{2}} \times b^{-\frac{1}{2}}}{b}.$$

2

(*b*) Express $\frac{4}{x} - \frac{3}{x-3}$, $x \neq 0$, $x \neq 3$, as a single fraction in its simplest form.

3

(*c*) Prove that

$$(\cos x° + \sin x°)^2 = 1 + 2 \sin x° \cos x°.$$

2

[END OF QUESTION PAPER]

[BLANK PAGE]

[BLANK PAGE]

X100/201

NATIONAL
QUALIFICATIONS
2003

WEDNESDAY, 21 MAY
1.30 PM – 2.15 PM

MATHEMATICS
INTERMEDIATE 2
Units 1, 2 and 3
Paper 1
(Non-calculator)

Read carefully

1 **You may NOT use a calculator.**

2 Full credit will be given only where the solution contains appropriate working.

3 Square-ruled paper is provided.

SCOTTISH
QUALIFICATIONS
AUTHORITY

FORMULAE LIST

The roots of $ax^2 + bx + c = 0$ are $x = \dfrac{-b \pm \sqrt{(b^2 - 4ac)}}{2a}$

Sine rule: $\dfrac{a}{\sin A} = \dfrac{b}{\sin B} = \dfrac{c}{\sin C}$

Cosine rule: $a^2 = b^2 + c^2 - 2bc \cos A$ or $\cos A = \dfrac{b^2 + c^2 - a^2}{2bc}$

Area of a triangle: $\text{Area} = \frac{1}{2}ab \sin C$

Volume of a sphere: $\text{Volume} = \frac{4}{3}\pi r^3$

Volume of a cone: $\text{Volume} = \frac{1}{3}\pi r^2 h$

Volume of a cylinder: $\text{Volume} = \pi r^2 h$

Standard deviation: $s = \sqrt{\dfrac{\sum (x - \bar{x})^2}{n - 1}} = \sqrt{\dfrac{\sum x^2 - (\sum x)^2 / n}{n - 1}}$, where n is the sample size.

ALL questions should be attempted.

Marks

1. Multiply out the brackets and collect like terms.

    $(2a - b)(3a + 2b)$

    2

2. Two spinners are used in an experiment.

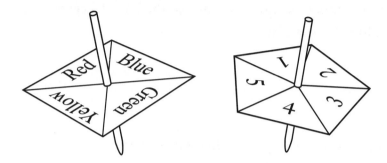

    The table below shows some of the possible outcomes when both spinners are spun and allowed to come to rest.

	1	2	3	4	5
Red	R,1	R,2			
Yellow	Y,1				
Blue	B,1				
Green	G,1				

   (a) Copy and complete the table.

   1

   (b) What is the probability that one spinner comes to rest on red and the other on an even number?

   1

[Turn over

Marks

3. The diagram shows a cone.

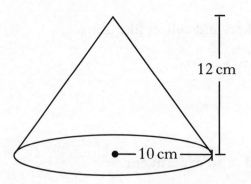

12 cm

10 cm

The height is 12 centimetres and the radius of the base 10 centimetres.

Calculate the volume of the cone.

Take π = 3·14.   **2**

4. A hotel books taxis from a company called QUICKCARS.

The receptionist notes the waiting time for every taxi ordered over a period of two weeks.

The times are recorded in the stem and leaf diagram shown below.

Waiting time (minutes)

```
0 | 6  7
1 | 2  3  4
2 | 5  6  9  9
3 | 2  5  7
4 | 2  4
```

n = 14   1 | 3 represents 13 minutes

(a) For the given data, calculate:

    (i) the median;   **1**

    (ii) the lower quartile;   **1**

    (iii) the upper quartile.   **1**

(b) Calculate the semi-interquartile range.   **1**

In another two week period, the hotel books taxis from a company called FASTCABS.

The semi-interquartile range for FASTCABS is found to be 2·5 minutes.

(c) Which company provides the more consistent service?

Give a reason for your answer.   **1**

Marks

5. Part of the graph of $y = a \sin bx°$ is shown in the diagram.

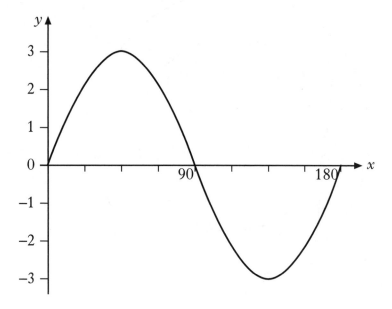

State the values of a and b.

2

6. (*a*) Express $\dfrac{\sqrt{40}}{\sqrt{2}}$ as a surd in its simplest form.

2

(*b*) Simplify $\dfrac{2x+2}{(x+1)^2}$.

2

[Turn over for Questions 7 and 8 on *Page six*

Marks

7.

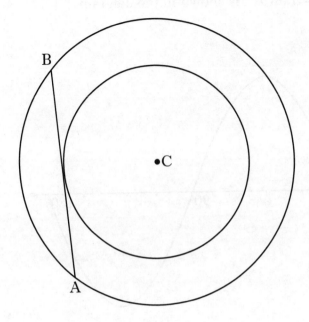

C is the centre of two concentric circles.

AB is a tangent to the smaller circle and a chord of the larger circle.

The radius of the smaller circle is 6 centimetres and the chord AB has length 16 centimetres.

Calculate the radius of the larger circle.

3

8. (*a*) Factorise $7 + 6x - x^2$.

2

(*b*) Hence write down the roots of the equation

$$7 + 6x - x^2 = 0.$$

1

(*c*) The graph of $y = 7 + 6x - x^2$ is shown in the diagram.

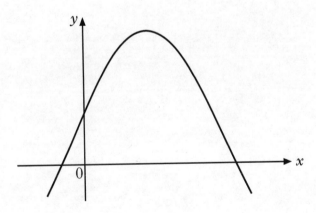

Find the coordinates of the turning point.

3

[*END OF QUESTION PAPER*]

X100/203

NATIONAL
QUALIFICATIONS
2003

WEDNESDAY, 21 MAY
2.35 PM – 4.05 PM

MATHEMATICS
INTERMEDIATE 2
Units 1, 2 and 3
Paper 2

Read carefully

1 **Calculators may be used in this paper.**

2 Full credit will be given only where the solution contains appropriate working.

3 Square-ruled paper is provided.

SCOTTISH
QUALIFICATIONS
AUTHORITY

FORMULAE LIST

The roots of $ax^2 + bx + c = 0$ are $x = \dfrac{-b \pm \sqrt{(b^2 - 4ac)}}{2a}$

Sine rule: $\dfrac{a}{\sin A} = \dfrac{b}{\sin B} = \dfrac{c}{\sin C}$

Cosine rule: $a^2 = b^2 + c^2 - 2bc \cos A$ or $\cos A = \dfrac{b^2 + c^2 - a^2}{2bc}$

Area of a triangle: $\text{Area} = \frac{1}{2} ab \sin C$

Volume of a sphere: $\text{Volume} = \frac{4}{3} \pi r^3$

Volume of a cone: $\text{Volume} = \frac{1}{3} \pi r^2 h$

Volume of a cylinder: $\text{Volume} = \pi r^2 h$

Standard deviation: $s = \sqrt{\dfrac{\sum (x - \bar{x})^2}{n - 1}} = \sqrt{\dfrac{\sum x^2 - (\sum x)^2 / n}{n - 1}}$, where n is the sample size.

ALL questions should be attempted.

Marks

1.

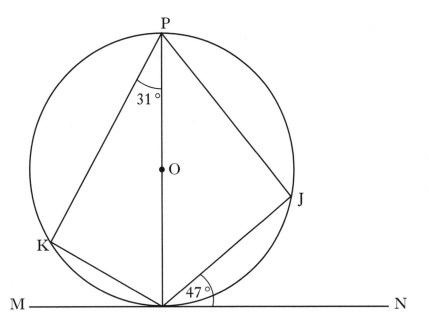

The tangent, MN, touches the circle, centre O, at L.

Angle JLN = 47°.

Angle KPL = 31°.

Find the size of angle KLJ.

3

2. A sample of shoppers was asked which brand of washing powder they preferred.

The responses are shown below.

Washing Powder	Frequency
Dazzle	250
Cyclo	375
Surfer	125
Cleano	250

Construct a pie chart to illustrate this information.

Show all your working.

3

[Turn over

Marks

3. Seats on flights from London to Edinburgh are sold at two prices, £30 and £50.

   On one flight a total of 130 seats was sold.

   Let x be the number of seats sold at £30 and y be the number of seats sold at £50.

   (a) Write down an equation in x and y which satisfies the above condition.　　1

   The sale of the seats on this flight totalled £6000.

   (b) Write down a second equation in x and y which satisfies this condition.　　1

   (c) How many seats were sold at each price?　　4

4. A bath contains 150 litres of water.

   Water is drained from the bath at a steady rate of 30 litres per minute.

   The graph of the volume, V litres, of water in the bath against the time, t minutes, is shown below.

   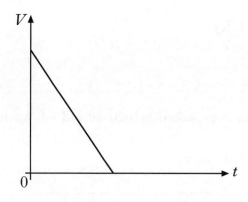

   Write down an equation connecting V and t.　　3

Marks

5. A gardener grows tomatoes in his greenhouse.

The temperature of the greenhouse, in degrees Celsius, is recorded every day at noon for one week.

17   22   25   16   21   16   16

(a) For the given temperatures, calculate:

   (i) the mean;

1

   (ii) the standard deviation.

3

  Show clearly all your working.

For best growth, the mean temperature should be $(20 \pm 5)°C$ and the standard deviation should be less than $5\,°C$.

(b) Are the conditions in the greenhouse likely to result in best growth?

  Explain clearly your answer.

2

[Turn over

6. A garden trough is in the shape of a prism.

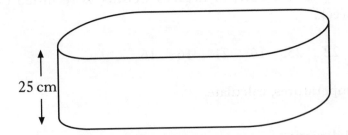

The height of the trough is 25 centimetres.
The cross-section of the trough consists of a rectangle and two semi-circles with measurements as shown.

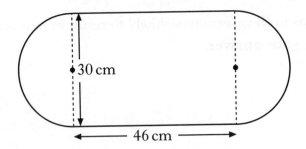

(a) Find the volume of the garden trough in cubic centimetres.
   Give your answer correct to two significant figures.　　　　4

A new design of garden trough is planned by the manufacturer.

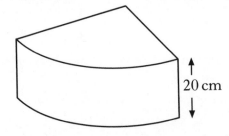

The height of the trough is 20 cm.
The uniform cross-section of this trough is a quarter of a circle.
The volume of the trough is 30 000 cm^3.

(b) Find the radius of the cross-section.　　　　3

Marks

7. Change the subject of the formula

$$y = ax^2 + c \qquad \text{to } x.$$

3

8. The diagram below shows a big wheel at a fairground.

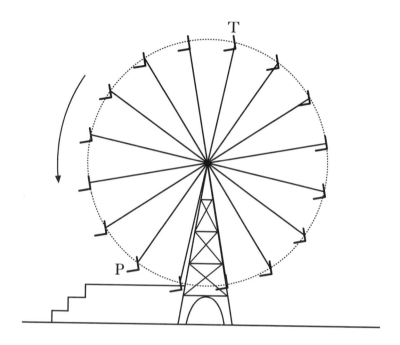

The wheel has sixteen chairs equally spaced on its circumference.

The radius of the wheel is 9 metres.

As the wheel rotates in an anticlockwise direction, find the distance a chair travels in moving from position T to position P in the diagram.

4

9. Solve the equation

$$2x^2 + 4x - 9 = 0,$$

giving the roots correct to one decimal place.

4

[Turn over for Questions 10 to 12 on *Page eight*

Marks

10. The sketch shows a parallelogram, PQRS.

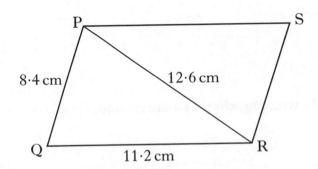

(a) Calculate the size of angle PQR.

Do not use a scale drawing.

3

(b) Calculate the area of the parallelogram.

3

11. (a) Express

$$a^{\frac{2}{3}}(a^{\frac{2}{3}} - a^{-\frac{2}{3}})$$

in its simplest form.

2

(b) Express

$$\frac{a}{x} - \frac{b}{y}, \qquad x \neq 0, \quad y \neq 0,$$

as a fraction in its simplest form.

2

12. (a) Solve the equation

$$2\tan x^\circ + 7 = 0, \qquad 0 \leq x < 360.$$

3

(b) Prove that

$$\sin^3 x^\circ + \sin x^\circ \cos^2 x^\circ = \sin x^\circ.$$

2

[END OF QUESTION PAPER]

[BLANK PAGE]

X100/201

| NATIONAL QUALIFICATIONS 2004 | FRIDAY, 21 MAY 1.00 PM – 1.45 PM | MATHEMATICS INTERMEDIATE 2 Units 1, 2 and 3 Paper 1 (Non-calculator) |

Read carefully

1 **You may <u>NOT</u> use a calculator.**

2 Full credit will be given only where the solution contains appropriate working.

3 Square-ruled paper is provided.

SCOTTISH
QUALIFICATIONS
AUTHORITY

©

FORMULAE LIST

The roots of $ax^2 + bx + c = 0$ are $x = \dfrac{-b \pm \sqrt{\left(b^2 - 4ac\right)}}{2a}$

Sine rule: $\dfrac{a}{\sin A} = \dfrac{b}{\sin B} = \dfrac{c}{\sin C}$

Cosine rule: $a^2 = b^2 + c^2 - 2bc \cos A$ or $\cos A = \dfrac{b^2 + c^2 - a^2}{2bc}$

Area of a triangle: Area $= \frac{1}{2}ab \sin C$

Volume of a sphere: Volume $= \frac{4}{3}\pi r^3$

Volume of a cone: Volume $= \frac{1}{3}\pi r^2 h$

Volume of a cylinder: Volume $= \pi r^2 h$

Standard deviation: $s = \sqrt{\dfrac{\sum (x - \bar{x})^2}{n-1}} = \sqrt{\dfrac{\sum x^2 - (\sum x)^2 / n}{n-1}}$, where n is the sample size.

Marks

ALL questions should be attempted.

1. In a class test, the following marks were recorded.

    | 5 | 9 | 10 | 4 | 5 | 5 | 6 | 10 | 5 | 8 |
    | 5 | 7 | 4 | 9 | 7 | 5 | 4 | 6 | 5 | 7 |

    (a) Construct a frequency table for the above data and add a cumulative frequency column.

    2

    (b) What is the probability that a student, chosen at random from this class, obtained a mark higher than 7?

    1

2.

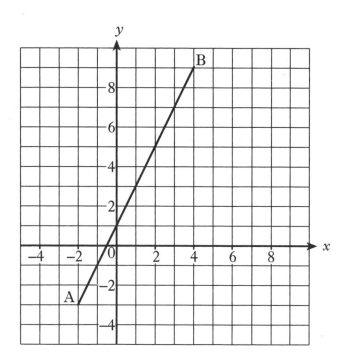

Find the equation of the straight line AB.

3

[Turn over

3.

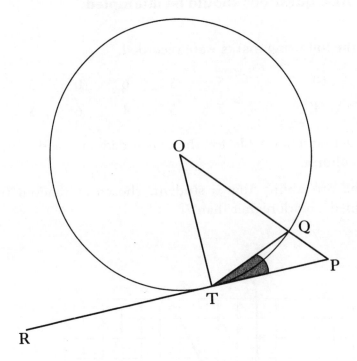

RP is a tangent to the circle, centre O, with a point of contact T.

The shaded angle PTQ = 24°.

Calculate the size of angle OPT.   **3**

4. The number of chocolates in each box from a sample of 25 boxes was counted.

The results are displayed in the dotplot below.

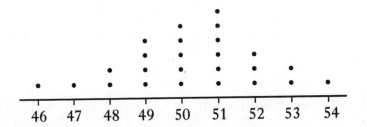

(*a*) For this sample find:

    (i)   the median;   **1**

    (ii) the lower quartile;   **1**

    (iii) the upper quartile.   **1**

(*b*) Use the data from this sample to construct a boxplot.   **2**

(*c*) In a second sample of boxes, the semi-interquartile range was 1·5.

Make an appropriate comment about the distribution of data in the two samples.   **2**

Marks

5. William Watson's Fast Foods use a logo based on parts of three identical parabolas.

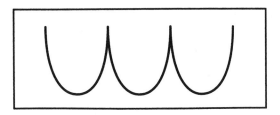

This logo is represented on the diagram below.

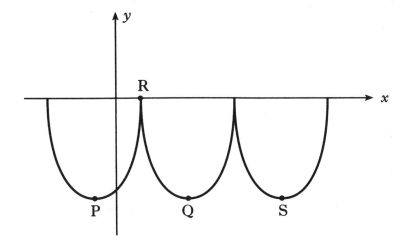

The first parabola has turning point P and equation $y = (x + 2)^2 - 16$.

(a) State the coordinates of P. **2**

(b) If R is the point (2, 0), find the coordinates of Q, the minimum turning point of the second parabola. **1**

(c) Find the equation of the parabola with turning point S. **2**

[Turn over for Question 6 on *Page six*

Marks

6. (*a*) Part of the graph of $y = b\cos ax°$ is shown in the diagram.

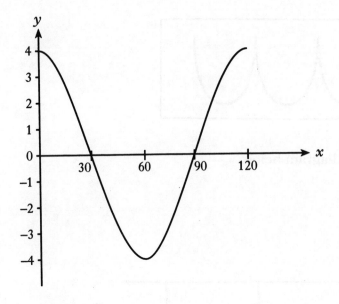

State the values of a and b

2

(*b*) Express $\sqrt{12} + 5\sqrt{3} - \sqrt{27}$ as a surd in its simplest form.

3

[END OF QUESTION PAPER]

X100/203

NATIONAL
QUALIFICATIONS
2004

FRIDAY, 21 MAY
2.05 PM – 3.35 PM

MATHEMATICS
INTERMEDIATE 2
Units 1, 2 and 3
Paper 2

Read carefully

1 **Calculators may be used in this paper.**

2 Full credit will be given only where the solution contains appropriate working.

3 Square-ruled paper is provided.

SCOTTISH
QUALIFICATIONS
AUTHORITY

©

FORMULAE LIST

The roots of $ax^2 + bx + c = 0$ are $x = \dfrac{-b \pm \sqrt{(b^2 - 4ac)}}{2a}$

Sine rule: $\quad \dfrac{a}{\sin A} = \dfrac{b}{\sin B} = \dfrac{c}{\sin C}$

Cosine rule: $\quad a^2 = b^2 + c^2 - 2bc \cos A \ $ or $ \ \cos A = \dfrac{b^2 + c^2 - a^2}{2bc}$

Area of a triangle: $\quad$ Area $= \frac{1}{2} ab \sin C$

Volume of a sphere: $\quad$ Volume $= \frac{4}{3} \pi r^3$

Volume of a cone: $\quad$ Volume $= \frac{1}{3} \pi r^2 h$

Volume of a cylinder: $\quad$ Volume $= \pi r^2 h$

Standard deviation: $\quad s = \sqrt{\dfrac{\sum (x - \bar{x})^2}{n - 1}} = \sqrt{\dfrac{\sum x^2 - (\sum x)^2 / n}{n - 1}}$, where n is the sample size.

ALL questions should be attempted.

Marks

1. The average Scottish house price is £77 900.

   The average price is expected to rise by 2·5% per month. What will the average Scottish house price be in 3 months?

   Give your answer correct to three significant figures.

   3

2. The heights, in millimetres, of six seedlings are given below.

   $$15 \quad 18 \quad 14 \quad 17 \quad 16 \quad 19$$

   (*a*) Calculate:

   (i) the mean;

   1

   (ii) the standard deviation;

   3

   of these heights.

   Show clearly all your working.

   (*b*) Later the same six seedlings are measured again.

   Each has grown by 4 millimetres.

   State:

   (i) the mean;

   1

   (ii) the standard deviation;

   1

   of the new heights.

3. (*a*) Multiply out the brackets and collect like terms.

   $$5x + (x - 4)(3x + 1)$$

   3

   (*b*) Factorise

   $$3x^2 - 7x + 2.$$

   2

   [Turn over

Marks

4.

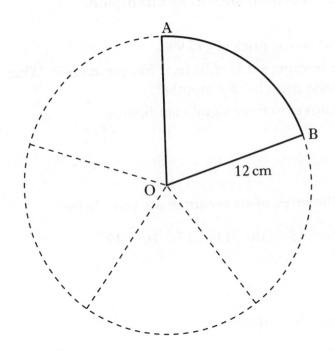

A circle, with centre O and radius 12 centimetres, is cut into 5 equal sectors. Calculate the perimeter of sector OAB.

3

5. A sports centre charges different entrance fees for adults and children.

   (*a*) One evening 14 adults and 4 children visited the sports centre. The total collected in entrance fees was £55·00.

     Let £x be the adult's entrance fee and £y be the child's entrance fee.

     Write down an equation in x and y which represents the above condition.

1

   (*b*) The following evening 13 adults and 6 children visited the sports centre. The total collected in entrance fees was £54·50.

     Write down a second equation in x and y which represents the above condition.

1

   (*c*) Calculate the entrance fee for an adult and the entrance fee for a child.

4

Marks

6. Solve the equation $2x^2 + 7x - 3 = 0$, giving the roots correct to one decimal place.

4

7. A garden, in the shape of a quadrilateral, is represented in the diagram below.

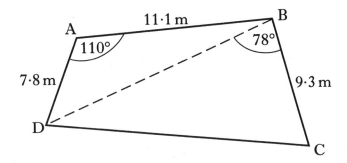

Calculate:

(*a*) the length of the diagonal BD;

   Do not use a scale drawing

3

(*b*) the area of the garden.

4

[Turn over

8. The diagram shows an L-shaped metal plate.

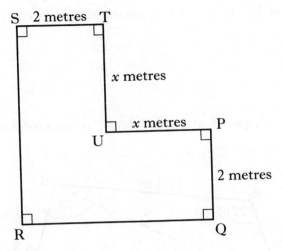

PQ = ST = 2 metres

TU = UP = x metres

(*a*) Show that the area, A square metres, of the metal plate is given by

$$A = 4x + 4.$$

2

(*b*) The area of the metal plate is 18 square metres.

Find x.

1

9. Perfecto Ice Cream is sold in cones and cylindrical tubs with measurements as shown below.

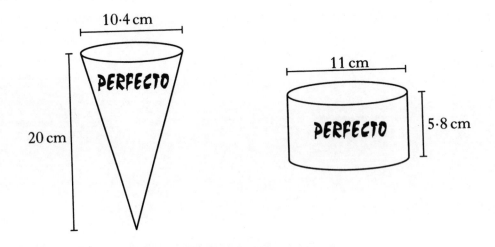

Both the cone and the tub of ice cream cost the same.

Which container of ice cream is better value for money?

Give a reason for your answer.

5

Marks

10. Solve the following equation for $0 \le x \le 360$.

$$7 \sin x° - 3 = 0$$

3

11. (*a*) Express $\dfrac{4}{x + 3} + \dfrac{3}{x}$, $x \ne -3$, $x \ne 0$,

as a single fraction in its simplest form.

3

(*b*) Change the subject of the formula $m = \dfrac{3x + 2y}{p}$ to x.

3

(*c*) Simplify $\dfrac{3a^5 \times 2a}{a^2}$

3

[END OF QUESTION PAPER]

[BLANK PAGE]

[BLANK PAGE]

X100/201

NATIONAL QUALIFICATIONS 2005	FRIDAY, 20 MAY 1.00 PM – 1.45 PM	**MATHEMATICS** INTERMEDIATE 2 Units 1, 2 and 3 Paper 1 (Non-calculator)

Read carefully

1   **You may NOT use a calculator.**

2   Full credit will be given only where the solution contains appropriate working.

3   Square-ruled paper is provided.

SCOTTISH
QUALIFICATIONS
AUTHORITY

©

FORMULAE LIST

The roots of $ax^2 + bx + c = 0$ are $x = \dfrac{-b \pm \sqrt{(b^2 - 4ac)}}{2a}$

Sine rule: $\dfrac{a}{\sin A} = \dfrac{b}{\sin B} = \dfrac{c}{\sin C}$

Cosine rule: $a^2 = b^2 + c^2 - 2bc \cos A$ or $\cos A = \dfrac{b^2 + c^2 - a^2}{2bc}$

Area of a triangle: Area $= \frac{1}{2}ab \sin C$

Volume of a sphere: Volume $= \frac{4}{3}\pi r^3$

Volume of a cone: Volume $= \frac{1}{3}\pi r^2 h$

Volume of a cylinder: Volume $= \pi r^2 h$

Standard deviation: $s = \sqrt{\dfrac{\sum(x - \bar{x})^2}{n - 1}} = \sqrt{\dfrac{\sum x^2 - (\sum x)^2 / n}{n - 1}}$, where n is the sample size.

Marks

ALL questions should be attempted.

1. The stem and leaf diagram below shows the heights of a group of children.

$$
\begin{array}{c|ccccccc}
12 & 1 & 2 & 4 & 5 & 9 \\
13 & 0 & 0 & 1 & 5 & 7 & 8 \\
14 & 0 & 2 & 8 & 9 \\
15 & 1 & 1 & 2 \\
\end{array}
$$

n = 18   12 | 1 represents 121 centimetres

What is the probability that a child chosen at random from this group has a height less than 130 centimetres?   **1**

2.

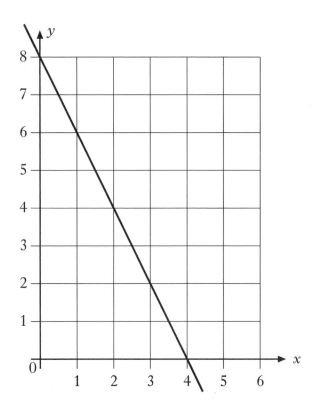

(a) Find the equation of the straight line shown in the diagram.   **3**

(b) Find the coordinates of the point where the line $y = 2x$ meets this line.   **2**

3. (a) Multiply out the brackets and collect like terms.

$$(4x + 2)(x - 5) + 3x$$ **3**

(b) Factorise

$$2p^2 - 5p - 12.$$ **2**

Marks

4. For a group of freezers in a shop, the volume, in litres, of each one is listed below.

$$78 \quad 81 \quad 91 \quad 75 \quad 85 \quad 83 \quad 84 \quad 78$$

(a) For the given data, calculate:

   (i) the median;   **1**

   (ii) the lower quartile;   **1**

   (iii) the upper quartile.   **1**

One of the numbers from the above list was accidentally missed out. A boxplot was then drawn and is shown below.

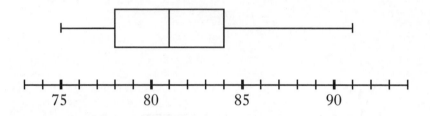

(b) Which number was missed out?

   Give a reason for your answer.   **2**

5. Simplify

$$k^8 \times (k^2)^{-3}.$$ **2**

6. Given that

$$\tan 45° = 1,$$

what is the value of $\tan 135°$?   **1**

7. Sketch the graph of

$$y = \sin 2x°, \quad 0 \le x \le 360.$$ **3**

Marks

8. A rectangle has length $(x + 2)$ centimetres and breadth x centimetres.

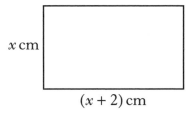

x cm

$(x + 2)$ cm

(*a*) Write down an expression for the area of the rectangle.   **1**

A square has length $(x + 1)$ centimetres.

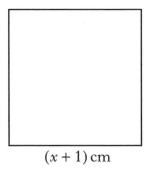

$(x + 1)$ cm

(*b*) The area of the square above is greater than the area of the rectangle. By how much is it greater?   **2**

[Turn over for Question 9 on *Page six*

Marks

9. The diagram below shows part of the graph of $y = 36 - (x - 2)^2$.

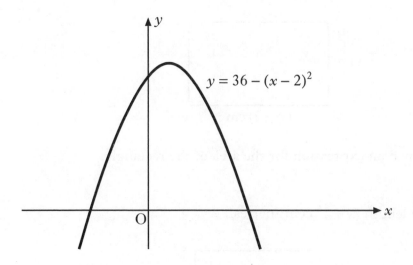

(*a*) State the coordinates of the maximum turning point.   2

(*b*) State the equation of the axis of symmetry.   1

The line $y = 20$ is drawn.
It cuts the graph of $y = 36 - (x - 2)^2$ at R and S as shown below.

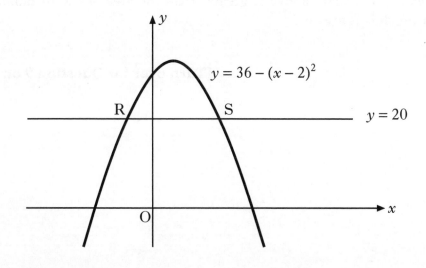

(*c*) S is the point (6, 20). Find the coordinates of R.   2

[END OF QUESTION PAPER]

X100/203

NATIONAL
QUALIFICATIONS
2005

FRIDAY, 20 MAY
2.05 PM – 3.35 PM

MATHEMATICS
INTERMEDIATE 2
Units 1, 2 and 3
Paper 2

Read carefully

1 **Calculators may be used in this paper.**

2 Full credit will be given only where the solution contains appropriate working.

3 Square-ruled paper is provided.

FORMULAE LIST

The roots of $ax^2 + bx + c = 0$ are $x = \dfrac{-b \pm \sqrt{(b^2 - 4ac)}}{2a}$

Sine rule: $\dfrac{a}{\sin A} = \dfrac{b}{\sin B} = \dfrac{c}{\sin C}$

Cosine rule: $a^2 = b^2 + c^2 - 2bc \cos A$ or $\cos A = \dfrac{b^2 + c^2 - a^2}{2bc}$

Area of a triangle: $\text{Area} = \tfrac{1}{2} ab \sin C$

Volume of a sphere: $\text{Volume} = \tfrac{4}{3} \pi r^3$

Volume of a cone: $\text{Volume} = \tfrac{1}{3} \pi r^2 h$

Volume of a cylinder: $\text{Volume} = \pi r^2 h$

Standard deviation: $s = \sqrt{\dfrac{\sum(x - \bar{x})^2}{n-1}} = \sqrt{\dfrac{\sum x^2 - (\sum x)^2 / n}{n-1}}$, where n is the sample size.

ALL questions should be attempted.

Marks

1. In the evening, the temperature in a greenhouse drops by 4% per hour.

   At 8 pm the temperature is 28 ° Celsius.

   What will the temperature be at 11 pm?   **3**

2. In a bakery, a sample of six fruit loaves is selected and the weights, in grams, are recorded.

   $$395 \quad 400 \quad 408 \quad 390 \quad 405 \quad 402$$

   For the above data the mean is found to be 400 grams.

   (*a*) Calculate the standard deviation.

   Show clearly all your working.   **3**

   (*b*) New methods are introduced to ensure more consistent weights.

   Another sample is then taken and the mean and standard deviation found to be 400 grams and 5·8 grams respectively.

   Are the new methods successful?

   Give a reason for your answer.   **1**

3. A straight line has equation $3y = 12 - 4x$.

   Find the coordinates of the point where it crosses the x-axis.   **2**

[Turn over

Marks

4. A jeweller uses two different arrangements of beads and pearls.

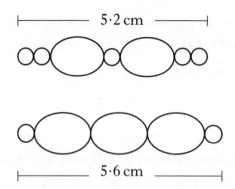

The first arrangement consists of 2 beads and 5 pearls and has an overall length of 5·2 centimetres.

The second arrangement consists of 3 beads and 2 pearls and has an overall length of 5·6 centimetres.

Find the length of **one** bead and the length of **one** pearl.　　　**6**

5. The diagram below shows a sector of a circle, centre C.

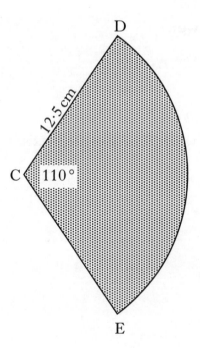

The radius of the circle is 12·5 centimetres and angle DCE is 110°.

Calculate the area of the sector CDE.　　　**3**

Marks

6. In the diagram below three towns, Holton, Kilter and Malbrigg are represented by the points H, K and M respectively.

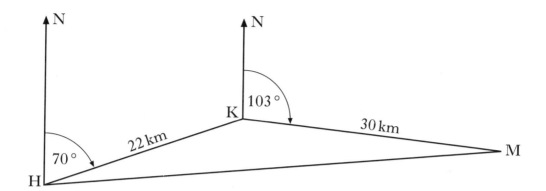

A helicopter flies from Holton for 22 kilometres on a bearing of 070° to Kilter. It then flies from Kilter for 30 kilometres on a bearing of 103° to Malbrigg. The helicopter then returns directly to Holton.

(*a*) (i) Calculate the size of angle HKM.   **1**

    (ii) Calculate the total distance travelled by the helicopter.   **3**

  Do not use a scale drawing.

(*b*) A climber is reported missing somewhere in the triangle represented by HKM in the diagram.

  Calculate the area of this triangle.   **2**

7. A pharmaceutical company makes vitamin pills in the shape of spheres of radius 0·5 centimetres.

(*a*) Calculate the volume of **one** pill.

  Give your answer correct to two significant figures.   **3**

The company decides to change the shape of each pill to a cylinder.

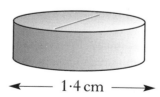

1·4 cm

(*b*) The new pill has the **same** volume as the original and its diameter is 1·4 centimetres.

  Calculate the height of the new pill.   **3**

[Turn over

Marks

8. Solve the equation

$$4x^2 - 7x + 1 = 0$$

giving the roots correct to one decimal place.

4

9. Points A, B and C lie on the circumference of a circle, centre O.

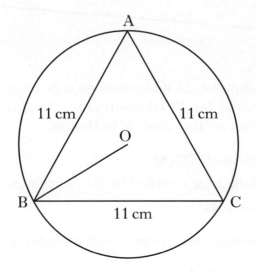

Triangle ABC is equilateral with sides of length 11 centimetres as shown in the diagram.

(*a*) Write down the size of angle OBC.

1

(*b*) Calculate the length of the radius OB.

3

Marks

10. (*a*) Express $\dfrac{7}{\sqrt{2}}$ as a fraction with a rational denominator.   **2**

(*b*) Express $\dfrac{a}{b} \times \dfrac{3b}{a^2}$ as a fraction in its simplest form.   **2**

(*c*) Change the subject of the formula

$$p = q + 2r^2 \quad \text{to } r.$$ **3**

11. (*a*) Solve the equation

$$7 \cos x^\circ - 5 = 0, \qquad 0 \le x < 360.$$ **3**

(*b*) Simplify

$$\tan x^\circ \cos x^\circ.$$ **2**

[END OF QUESTION PAPER]

[BLANK PAGE]

[BLANK PAGE]

[BLANK PAGE]

Mathematics Intermediate 2
Units 1, 2 and 3
Paper 2 Winter Diet 2002

1. $y = \sqrt{x + m}$

2. $22 \cdot 6 \, m^2$

3. (a) 2%

   (b) £86,700 $\times 1 \cdot 02^3$ = £92,000

4. $200 \, m^3$

5. (a) $3x + 2y = 8 \cdot 60$

   (b) $5x + 3y = 13 \cdot 60$

   (c) cost of ticket for ghost
   train = £1·40
   cost of ticket for sky
   ride = £2·20

6. (a) £18·20

   (b) 2·4 (disregard rounding)

   (c) (i) £23·20

       (ii) standard deviation is the
       same (2·4)

7. (a) angle PQR = $68 \cdot 9°$

   (b) $62 \cdot 4 \, m^2$

8. $p = -0 \cdot 3$ and $1 \cdot 8$

9. 21·5 metres

10. (a) $2x^3 - 7x^2 - 11x + 6$

    (b) $(2x - 9)(x + 1)$

11. 26·6, 206·6

12. (a) b

    (b) $\dfrac{x - 12}{x(x - 3)}$ **or** $\dfrac{x - 12}{x^2 - 3x}$

    (c) $(\cos x° + \sin x°)^2$

    $= \cos^2 x° + 2\sin x° \cos x° + \sin^2 x°$

    $= \sin^2 x° + \cos^2 x° + 2\sin x° \cos x°$

    $= 1 + 2\sin x° \cos x°$

    (since $\sin^2 x° + \cos^2 x° = 1$)

Mathematics Intermediate 2
Units 1, 2 and 3
Paper 1 (Non-calculator)
2003

1. $6a^2 + ab - 2b^2$

2. (a)

	1	2	3	4	5
Red	R,1	R,2	R,3	R,4	R,5
Yellow	Y,1	Y,2	Y,3	Y,4	Y,5
Blue	B,1	B,2	B,3	B,4	B,5
Green	G,1	G,2	G,3	G,4	G,5

   (b) $\dfrac{2}{20}$ (or equivalent)

3. $1256 \, cm^3$

4. (a) (i) 27·5

       (ii) 13

       (iii) 35

   (b) 11

   (c) FASTCABS with valid reason,
   e.g.

   Fastcabs because their semi-
   interquartile range is much
   smaller

   Fastcabs because their waiting
   times are less spread out

5. $a = 3, b = 2$

6. (a) $2\sqrt{5}$

   (b) $\dfrac{2}{x + 1}$

7. 10 cm

8. (a) $(7 - x)(1 + x)$

   (b) 7, −1

   (c) (3, 16)

Mathematics Intermediate 2
Units 1, 2 and 3
Paper 2
2003

1. $102°$

2.

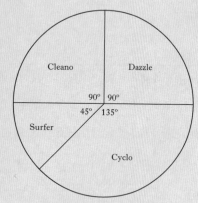

3. (a) $x + y = 130$

    (b) $30x + 50y = 6000$

    (c) 25 seats sold at £30, 105 seats sold at £50

4. $V = -30t + 150$

5. (a) (i) $x = 19$

       (ii) $s = 3·65$

    (b) Yes, with reasons covering both conditions

6. (a) $52\,000\ \text{cm}^3$

    (b) $43·7\ \text{cm}$

7. $x = \sqrt{\dfrac{y-c}{a}}$

8. $24·7\ \text{m}$

9. $x = -3·3,\ x = 1·3$

10. (a) $PQR = 78·6°$

    (b) $92·2\ \text{cm}^2$

11. (a) $a^{4/3} - 1$

    (b) $\dfrac{ay - bx}{xy}$

12. (a) $x = 105·9$ and $285·9$

    (b) Proof, e.g.
$$\sin x°(\sin^2 x° + \cos^2 x°)$$
$$= \sin x° \times 1$$
$$= \sin x°$$

Mathematics Intermediate 2
Units 1, 2 and 3
Paper 1 (Non-calculator)
2004

1. (a)

frequency	cumulative frequency
3	3
7	10
2	12
3	15
1	16
2	18
2	20

    (b) $\dfrac{5}{20}$ or equivalent

2. $y = 2x + 1$

3. $42°$

4. (a) (i) $Q_2 = 50$

       (ii) $Q_1 = 49$

       (iii) $Q_3 = 51·5$

    (b)

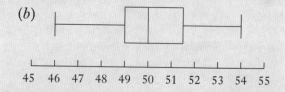

     45 46 47 48 49 50 51 52 53 54 55

    (c) The SIQR in first sample was $1·25$ which is less than $1·5$ so number of chocolates in each box in first sample is more consistent.

5. (a) $P(-2, -16)$

    (b) $Q(6, -16)$

    (c) $y = (x - 14)^2 - 16$

6. (a) $a = 3,\ b = 4$

    (b) $4\sqrt{3}$

Mathematics Intermediate 2
Units 1, 2 and 3
Paper 1 (Non-calculator)
Winter Diet 2002

1. (*a*) $F = 2c + 5$

   (*b*) mark = 29 (evidence of substitution must be shown)

2. angle OPT = 40°

3. (*a*) (i)   median = 171

       (ii) $Q_1 = 162$, $Q_3 = 174{\cdot}5$

(*b*)

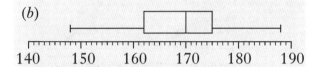

| 140 | 150 | 160 | 170 | 180 | 190 |

(*c*) Two appropriate comments which involve **different** aspects of the box plots e.g.

   1. The range is greater for females

      or

      The males are less spread out

      and

   2. The median of the males is greater than the median of the females

      or

      on average the males are taller

4. $a = -3$

5. (*a*) $2\sqrt{5}$

   (*b*) 36

6. $b = 60$

7. (*a*) length = $9 + x$
          breadth = $5 + x$

7. (*b*) Area $= (9 + x)(5 + x)$
              $= 45 + 9x + 5x + x^2$
              $= x^2 + 14x + 45$

   (*c*) $x^2 + 14x + 45 = 77$
       $x^2 + 14x - 32 = 0$
       $(x + 16)(x - 2) = 0$
       $x = -16$, $x = 2$
       Width is 2 metres

Pocket answer section for
SQA Mathematics Intermediate 2: Units 1, 2 and 3
2002, 2002 Winter Diet, 2003, 2004 and 2005

Mathematics Intermediate 2
Units 1, 2 and 3
Paper 1 (Non-calculator)
2002

1. (a)

	Frequency	Cumulative Frequency
70	2	2
71	3	5
72	3	8
73	3	11
74	2	13
75	2	15
76	1	16

   (b) $\frac{5}{16}$

2. $y = \frac{5}{2}x + 5$ or equivalent

3. 120, 240

4. $x^3 + x^2 - 13x + 3$

5. (a) $Q_1 = 1\cdot5$, median = 3, $Q_3 = 4$

    (b)
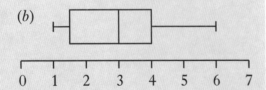

   (c) more spread out or higher median

6. (a) (1, −16)

    (b) $x = 1$

    (c) AB = 8

7. (a) $\sqrt{5}$

    (b) $\frac{1+x}{x^2}$

Mathematics Intermediate 2
Units 1, 2 and 3
Paper 2
2002

1. $5438 \, \text{m}^2$

2. $x = 3$, $y = -1$

3. (a) $\bar{x} = 73$ and s.d. $= 10\cdot5$

    (b) same mean, but prices have greater spread in local shops

4. 82°

5. (a) (i) $3y\,(y-2)$

        (ii) $(y+3)\,(y-2)$

    (b) $\frac{3y}{y+3}$

6. $2000 \, \text{cm}^3$

7. $x = -1\cdot8$ and $0\cdot3$

8. 23·3 metres

9. 0·5 metres

10. 3 years

11. (a) $3x$

     (b) $p = \frac{r-2t}{3}$

12. (a) 10·868 m

     (b) 30 and 150 seconds

Mathematics Intermediate 2:
Units 1, 2 and 3
Paper 2
2005

1. 24·8°

2. (a) 6·6

   (b) YES, because new s < 6·6

3. (3,0)

4. bead 1·6cm, pearl 0·4cm

5. 149·9cm²

6. (a) (i) 147°
   (ii) 101·9km

   (b) 179·7km²

7. (a) 0·52cm³

   (b) 0·34cm

8. $x = 1·6$ or $0·2$

9. (a) 30°

   (b) 6·35cm

10. (a) $\dfrac{7\sqrt{2}}{2}$

    (b) $\dfrac{3}{a}$

    (c) $r=\sqrt{\dfrac{p-q}{2}}$

11. (a) 44.4° and 315·6°

    (b) $\sin x$

Mathematics Intermediate 2
Units 1, 2 and 3
Paper 2
2004 contd.

11. (a) $\dfrac{7x + 9}{x(x + 3)}$ OR $\dfrac{7x + 9}{x^2 + 3x}$

(b) $x = \dfrac{mp - 2y}{3}$

Working
1. $mp = 3x + 2y$
2. $3x = mp - 2y$
3. $x = \dfrac{mp - 2y}{3}$

(c) $6a^4$

Working
1. $\dfrac{3 \times 2a^6}{a^2}$
2. $\dfrac{6a^6}{a^2}$
3. $6a^4$

Mathematics Intermediate 2:
Units 1, 2 and 3
Paper 1 (Non-calculator)
2005

1. $\dfrac{5}{18}$

2. (a) $y = -2x + 8$

(b) $(2,4)$

3. (a) $4x^2 - 15x - 10$

(b) $(2p + 3)(p - 4)$

4. (a) (i) 82
   (ii) 78
   (iii) 84.5

(b) 85
   A valid reason could be a revised
   list indicating quartiles, e.g.
   75 78 78 81 83 84 91

5. k^2

6. -1

7.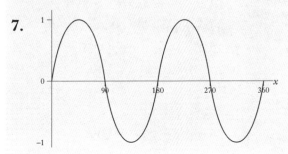

8. (a) $x(x + 2)$

(b) 1 cm^2

9. (a) $(2,36)$

(b) $x = 2$

(c) $(-2,20)$

Mathematics Intermediate 2
Units 1, 2 and 3
Paper 2
2004

1. £83 900

2. (a) (i) $x = 16\cdot5$
   (ii) $s = 1\cdot87$

   (b) (i) $x = 20\cdot5$
   (ii) $s = 1\cdot87$

3. (a) $3x^2 - 6x - 4$

   (b) $(3x - 1)(x - 2)$

4. $39\cdot1$ cm

5. (a) $14x + 4y = 55\cdot00$

   (b) $13x + 6y = 54\cdot50$

   (c) Entrance fee for adult is £3·50
   Entrance fee for child is £1·50

   Method 1 – Working
   1. $14x + 4y \quad = 55\cdot00 \quad (\times3)$
   $13x + 6y \quad = 54\cdot00 \quad (\times2)$

   Match the 'y's
   2. $42x + 12y \quad = 165$
   $26x + 12y \quad = 109$

   Subtract
   3. $42x + 12y \quad = 165$
   $26x + 12y \quad = 109$
   $16x \quad = 56$

   4. $x = 3\cdot5$
   5. $14x + 14y = 55$
   $(14\times3\cdot5) + 4y = 55$
   $49 + 4y = 55$
   $4y = 55 - 49$
   6. $y = 1\cdot5$

   Method 2
   Use equations for parts (a) and (b) to generate graphs to find x & y.

6. $x = -3\cdot9, x = 0\cdot4$

Method 1
   Use quadratic formula.
   1. $\dfrac{-7\pm\sqrt{7^2 - 4(2)(-3)}}{2(2)}$

   2. $b^2 - 4ac = 73$
   3. $-3\cdot9, 0\cdot4$

Method 2 – possible graphical solution
Draw graph of $y = 2x^2 + 7x - 3$, and indicate positions of roots:

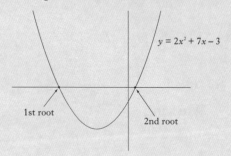

7. (a) $15\cdot6$ metres

   Working
   $11\cdot1^2 + 7\cdot8^2 - 11\cdot1 \times 7\cdot8 \times \cos 110$

   (b) $111\cdot6$ square metres

   Working
   1. $\frac{1}{2} \times 11\cdot1 \times 7\cdot8 \times \sin 110$
   2. $\frac{1}{2} \times 9\cdot3 \times$ answer to part (a) $\times \sin 78$

8. (a) 1. $1.2(x + 2) + 2x$ OR $2x + 2x + 4$
   (or equivalent)
   2. Area $= 4x + 4$

   (b) $x = 3\cdot5$

9. The cone is better value because it contains more ice cream.

   Working
   1. $V = \frac{1}{3} \times \pi \times 5\cdot2^2 \times 20$ (cone)
   2. $V = \pi \times 5\cdot5^2 \times 5\cdot8$ (tub)
   3. $566\cdot3$ cm³ (cone)
   $551\cdot2$ cm³ (tub)

10. $x = 25\cdot4$ and $154\cdot6$

   Working
   $\sin x = \frac{3}{7}$